Morse Code
for Radio Amateurs

9th edition

Roger Cooke, G3LDI

CONTENTS

Radio Society of Great Britain

Published by the Radio Society of Great Britain, Cranborne Road, Potters Bar, Herts EN6 3JE.

First published 2006

ISBN 1-905086-17-2
EAN 978-1905-086-17-7

Publisher's note

The opinions expressed in this book are those of the author and not necessarily those of the RSGB. While the information presented is believed to be correct, the author, publisher and their agents cannot accept responsibility for consequences arising from any inaccuracies or omissions.

Typography and design: Mike Dennison, G3XDV, Emdee Publishing
Cover design: Dorotea Vizer
Production: Mark Allgar, M1MPA

Printed in Great Britain by Latimer Trend of Plymouth

Cover image of a Kent Morse key was kindly supplied by Waters & Stanton PLC

1

The Morse Code and its History

MORSE CODE is a method of representing letters, numbers and punctuation marks by means of a code signal sent intermittently. It was developed by Samuel Morse (**Fig 1.1**).

Morse code is an early form of digital communications, something that is common-place today. However, unlike modern binary digital codes that use just two states (commonly represented as 1 and 0), it uses five: dot (·), dash (-), short gap (between each letter), medium gap (between words) and a long gap (between phrases or sentences, not often applied). Of course, Morse code is also a binary code, in that it is based on only two states -- on and off. Both views (dits/dahs/spaces versus on and off) are correct.

Samuel Morse was born in 1791, at Charlestown, Massachusetts. His father was a Congregational minister and a scholar of high standing, who, by careful management, was able to send his three sons to Yale College. Samuel (or Finley, as he was called by his family) attended at the age of fourteen and came under the influence of Benjamin Silliman, Professor of Chemistry, and of Jeremiah Day, Professor of Natural Philosophy, afterwards President of Yale College, whose teaching gave him impulses which in later years led to the invention of the telegraph. "Mr. Day's lectures are very interesting," the young student wrote home in 1809, "they are upon electricity; he has given us some

Fig 1.1: Alfred Vail (left) and Samuel Morse

very fine experiments, the whole class taking hold of hands form the circuit of communication and we all receive the shock apparently at the same moment." Electricity, however, was only an alluring study. It afforded no means of livelihood, and Morse had gifts as an artist; in fact, he earned a part of his college expenses painting miniatures at five dollars apiece. He decided, therefore, that art should be his vocation. However, he found it a very difficult and impecunious profession. He also taught art to supplement his income. One enthusiastic student who had little money for his art course paid $10 as a part payment for his course. This is what Samuel Morse said:

"This is my first meal for twenty-four hours. Strother, don't be an artist. It means beggary. Your life depends upon people who know nothing of your art and care nothing for you. A house dog lives better, and the very sensitiveness that stimulates an artist to work keeps him alive to suffering."

The Birth of the Telegraph

In the year 1836 Morse took into his confidence one of his colleagues in the University, Leonard D Gale, who assisted him greatly, in improving the telegraph invention, while the inventor himself formulated the rudiments of the telegraphic alphabet, or Morse code, as it is known today. At length all was ready for a test and the message flashed from transmitter to receiver. The telegraph was born, though only an infant as yet.

Morse concentrated more on his electrical experiments and on 2 September, 1837, a successful experiment was made with seventeen hundred feet of copper wire coiled around the room. This was in the presence of Alfred Vail, a student whose family owned the Speedwell Iron Works at Morristown, New Jersey. He at once took an interest in the invention and persuaded his father, Judge Stephen Vail, to advance money for experiments. Samuel Morse filed a petition for a patent in October and admitted his colleague Gale, as well as Alfred Vail, to partnership. Experiments followed at the Vail shops, all the partners working day and night in their enthusiasm. The apparatus was then brought to New York and gentlemen of the city were invited to the University to see it work before it left for Washington.

Fig 1.2: First telegraph model (c. 1835) made from an old artist canvas stretcher, home-made battery and wooden clockworks. Code was generated by the wooden arm riding across the metal saw tooth dies representing dots and dashes and printed out on the paper tape moved by the clockworks

The equipment was gradually improved and was demonstrated in 1837. He eventually applied for a patent in 1840. A line was set up between Baltimore and Washington and the first message, sent on 24 May, 1844, was: "What hath God wrought!" The original keyer, if you can call it that, or telegraph is shown in **Fig 1.2** but I don't think many modern amateurs would want one in the shack!

Morse's original code consisted of combinations of dots and dashes that represented numbers. Each

Fig 1.3: Morse/Vail telegraph register, 1844. This register was used to receive the message "What Hath God Wrought" on the experimental circuit between Washington DC, and Baltimore, Maryland

number represented a word. This required looking up the number in a book to find the word it represented. A telegraph key was then used to tap out the sequence of dots, dashes, and pauses that represented the number.

Although Morse invented the telegraph, he lacked technical expertise. He entered an agreement with Alfred Vail who built more practical equipment. Vail developed a system in which each letter or symbol is sent individually, using combinations of dots, dashes, and pauses. Morse and Vail agreed that Vail's method of representing individual symbols would be included in Morse's patent. This system, known as American Morse code, was the version that was used to transmit the first telegraph message. The receiving apparatus is shown in **Fig 1.3**.

The code may be transmitted as an audio tone, a steady radio signal switched on and off (only the carrier wave, or CW, also known as continuous wave), an electrical pulse down a telegraph wire, or as a mechanical or visual signal (such as a flashing light). It may also be used with a flash lamp for visible signaling. The International (or continental) Morse code is a simplified form generally used in radio telegraphy.

American Morse Code

American Morse differs from International Morse in eleven letters, in all of the numerals except the numeral '4', and in the punctuation code. The unit of the code is the dot, representing a very brief depression of the telegraph key. The dash represents a depression lasting three times as long as a dot. Between the depressions there is a pause equal in time to one dot, except in a few letters and signs, when there is a wait of two dots. The pause between letters in a word lasts as long as one dash, between words it lasts as long as two dashes. American Morse Code is sometimes referred to as 'Railroad Morse'. This code is virtually extinct and no longer in commercial use. This style of Morse was developed for land operators working over telegraph wire rather than via radio signals. It is most frequently seen today in railroad museums and American civil war re-enactments. There are still some old timers around who can use both. Dick Bendicksen, N7ZL, who died in 2006 aged 87, could send and receive at 40WPM using both code sets. That feat takes real skill and is quite extraordinary! To compare the two codes look at **Fig 1.4**.

International and American Morse

The Morse code was developed so that operators could translate the indentations marked on the paper tape into text messages. Initially, Morse had planned to transmit only numerals, and use a dictionary to look up each word according to the number

CHARACTER:	MORSE CODE AMERICAN MORSE	CONTINENTAL CODE INTERNATIONAL CODE
A	•—	•—
B	—•••	—•••
C	•• •	—•—•
D	—••	—••
E	•	•
F	•—•	••—•
G	——•	——•
H	••••	••••
I	••	••
J	—•—•	•———
K	—•—	—•—
L	——	•—••
M	——	——
N	—•	—•
O	• •	———
P	•••••	•——•
Q	••—•	——•—
R	• ••	•—•
S	•••	•••
T	—	—
U	••—	••—
V	•••—	•••—
W	•——	•——
X	•—••	—••—
Y	•• ••	—•——
Z	••• •	——••
1	•——•	•————
2	••—••	••———
3	•••—•	•••——
4	••••—	••••—
5	———	•••••
6	••••••	—••••
7	——••	——•••
8	—••••	———••
9	—••—	————•
0	——	—————
Period	••——••	•—•—•—
Comma	•—•—	——••——
Question	—••—•	••——••

Fig 1.4: Comparison between the early American Morse Code and the currently used International or Continental Code

which had been sent. However, the code was expanded to include letters and special characters, so it could be used for more complete messages. The shorter marks were called 'dots', and the longer ones 'dashes', and the letters most commonly used in the English language were assigned the shortest sequences.

In the original Morse telegraphs, the receiver's armature made a clicking noise as it moved into and out of position for marking the tape. Operators soon learned to directly read the clicks as the beginning and end of dots and dashes, meaning that it was no longer necessary to use the tape.

When Morse code was adapted to radio, the dots and dashes were normally sent as short and long tones. It was later found that people become more proficient at receiving Morse code when it is taught as a language that is heard, instead of one read from a page. To reflect the sound of Morse code, practitioners vocalise a dash as 'dah', and a dot as a 'dit'. When a dit is not the final element of a character, its sound is shortened to "di-" to maintain a better vocal rhythm. For instance, the letter F (••▬•) is 'di-di-dah-dit'.

This older style of code was developed to accommodate the way in which operators listened to Morse code sent to them. Rather than hearing tones from a speaker or headphones as we do now using International Morse Code, in the earliest days of telegraphy one would hear two clicks from a mechanical sounding device for each key movement. Pressing the key makes a click, and releasing the key makes a clack. Thus, each key movement, up or down was uniquely heard. In this mechanical sounder system, an A (•▬) would sound like: clickClack click - - - Clack. This is quite different from 'CW' code where beeps are heard for as long as the key is engaged.

Modern International Morse Code

The Modern International Morse Code was invented by Friedrich Clemens Gerke in 1848 and used for the telegraphy between Hamburg and Cuxhaven in Germany. After some minor changes in 1865 it was standardised at the International Telegraphy congress in Paris (1865), and later accepted by the ITU as International Morse Code.

International Morse code is still in use today, albeit not in the commercial world. The US Coast Guard stopped monitoring Morse code transmissions in 1995 when their use in sending distress calls had been almost entirely superseded by automated systems using satellite relay.

Morse has become almost exclusively the province of amateur radio operators. Until 2003 the International Telecommunications Union (ITU) mandated Morse code proficiency as part of the amateur radio licensing procedure throughout the world. In the amateur radio world, certain parts of the amateur radio bands are still reserved for transmission of Morse code signals only.

Since Morse relies on only an on-off keyed radio signal, it requires less complex equipment than other forms of radio communication, and it can be used in very high noise / low signal environments. It also requires less bandwidth than voice communications, typically 100-150Hz for hand-sent Morse. The extensive use of pro-signs, Q-codes, and restricted format of typical messages facilitates communication between amateur radio operators who do not share a common mother tongue and would have great difficulty in communicating using voice modes.

Outdated Mode?

Morse code's descendants and other newer fdigital modes have superseded Morse code, with the argument that newer 'high-tech' modes using computers are easier and faster to use. While it may be true that picking up a telephone is much simpler than having to train people in Morse code to transmit a message, there is the argument that technology is good only if technology is working. Morse code's simplicity has often been able to overcome technological problems. Military and civilian emergency communications specialists have pointed out that even if a high-tech voice, picture, or data-transmitting device is malfunctioning and can produce only pulsed static, it can still effectively relay a coherent message if it is used to produce and send Morse code

Theoretically a low keying speed occupies a small bandwidth, so that 12WPM will fit comfortably into a bandwidth of a few tens of Hertz. This does not quite work in practice. Even so, the ability to use narrow band filters - 250Hz typically - enables far more CW stations to occupy a given bandwidth, plus there is a greater tolerance of interference

Morse code is also popular among QRP operators for enabling very long distance, low-power communication. Readability can be sustained by trained operators even though the signal is only faintly audible. This level of 'penetration' is due to the fact that all transmitted energy is concentrated in a very small bandwidth making the use of a narrow receiver bandwidth practical. A narrow bandwidth receiver uses filters to exclude interference on frequencies close to the desired frequency. Concentrating the transmitted energy in a small bandwidth gives the signal a "spectral brightness" that is much higher than the average natural noise.

EME (Moonbounce) communications use Morse as the main operating mode. Handling extremely weak signals bounced off the surface of the Moon makes Morse an ideal mode.

High speed Morse is used in meteor scatter contacts on VHF. Reflection from the ionised trail of meteors lasts only a few seconds so speeds of several hundred words per minute are used. This is generated and decoded by computer, enabling communication to be established. Usually this is by pre-arranged skeds with each station transmitting on alternate 15-second intervals. However, it is still using Morse code.

Conversely, 136kHz operators achieve long distances by using extremely slow Morse with each dot several seconds (or even minutes) long, usually readable on a computer screen as the only way of deciphering the message.

Most HF DXpeditions use CW and some even favour it exclusively. If you chase DX or IOTA Islands, CW should be one of your prime modes. It is far from out-dated, but no longer form part of the licensing requirements in this modern age.

Forty years ago, most amateurs operated CW only because to operate telephony it was necessary to have a modulator. Not only that, but in those days it was mandatory to spend the first year on CW anyway. It is this generation of died in the wool CW operators that try to perpetuate the mode. There are still a large number of CW only contests too and we need to maintain interest in the mode and train the newcomers to become proficient operators. There is nothing quite like sitting on the CW end of 20 meters with headphones on, and tuning across conversations, chatting with friends and breaking the pile-up on a DXpedition! It's a fascinating and skillful world of its own.

2

Morse and the Radio Amateur

Fig 2.1: G3VZT running a pile-up on NFD

RADIO AMATEURS have been associated with the Morse code ever since the hobby started. The two are synonymous and the quintessential picture of a radio amateur is one of him sitting in front of radio equipment with headphones on pounding a Morse key.

However, the modern Morse code enthusiast will be found using an electronic keyer and paddle plus a computer, and the computer takes precedence these days in a contest, with the keyer and paddle serving as back-up.

This image has faded somewhat in the last few years, mainly due to newcomers to the hobby not having to learn Morse at all. Indeed a lot of newly licensed amateurs openly admit they have no interest at all in Morse and won't use it. However, listening on HF during a contest or to a pile-up calling a DX-pedition and you will find that it really is still as popular as it was.

You will still find plenty of amateurs licensed in the 1940s through to the 1980s using Morse, and some licensed after that of course. However, new licensees do not have to be able to read Morse. There is no longer a mandatory Morse test. In the UK, it has been replaced by a Morse assessment with the idea that it might perhaps encourage some to use the mode after having received a licence.

Some people have no desire at all to use Morse, but by not using the code they are virtually excluded from 40% of the HF bands. The RSGB Cumulative short contests have sparked a renewed interest in Morse.

In the United States until 1991, a demonstration of the ability to send and receive Morse code at 5 words per minute (WPM) was required to receive an FCC amateur radio licence. Demonstration of this ability is still required for the privilege to use the HF bands. Until 1999, proficiency at the 20WPM level was required to receive the highest level of amateur licence (Extra Class). As of 15 April, 2000, the FCC reduced the Extra Class requirement to 5WPM.

The World Radiocommunication Conference of 2003 (WRC-03) made optional the international Morse code requirement for amateur radio licensing. Although the requirement remains on the books in the US, Canada, and elsewhere, some countries are working to eliminate the requirement entirely.

Benefits of the Morse code

One of the main benefits of using Morse is that it makes the most efficient use of the bandwidth available, as discussed in Chapter 1. It is possible to have ten CW contacts in the space that one SSB signal would occupy. It is ideal for low power operation, and is a great mode for the beginner. It also enables the operator to use the hobby quietly without making a lot of noise in the house. Morse is the only mode that requires a practical skill by the operator to create the means of communication.

Morse is an art form in itself and operators take a great pride in their skill. It is possible to improve one's speed, regardless of what level has been achieved. Morse is also a therapeutic means of communications and can be quite relaxing.

Most amateurs that have taken the trouble to learn Morse can copy speeds of 30WPM-plus, in their heads. At that speed and ability it becomes conversation, with no need to write it down, other than jotting down notes on which to comment. At this stage it is pure pleasure. Although the traditional straight key is still used by many, at this speed it is not possible, and electronic keyers become essential pieces of equipment for the Morse enthusiast. Contests are operated by using computer software with keyboard control. Macros can be programmed and selected by the operator. I shall be covering this in a later chapter.

Morse code has a 21st century role as an assistive technology, helping people with a variety of disabilities to communicate. Morse can be sent by someone with severe motion disability, as long as they have some minimal motor control. In some cases this means alternately blowing into and sucking on a plastic tube ('puff and sip' interface). People with severe sensory disabilities (eg deaf and blind) can receive Morse through a skin buzzer. Products are available that allow a computer operating system to be controlled by Morse code, allowing the user access to the Internet and electronic mail.

Learning the Code

Some people make the mistake of thinking that the first step to learning Morse is to go out and buy a Morse key. This is the wrong approach. First of all, the alphabet and numbers 0-9 have to be learned parrot fashion. I am a firm believ-

Morse Code Table

A	•–	J	•–––	S	•••	1	•––––	/	–••–•
B	–•••	K	–•–	T	–	2	••–––	.	•–•–•–
C	–•–•	L	•–••	U	••–	3	•••––	,	––••––
D	–••	M	––	V	•••–	4	••••–	?	••––••
E	•	N	–•	W	•––	5	•••••		
F	••–•	O	–––	X	–••–	6	–••••		
G	––•	P	•––•	Y	–•––	7	––•••		
H	••••	Q	––•–	Z	––••	8	–––••		
I	••	R	•–•			9	––––•		
						0	–––––		

Procedural Signals (also called ProSigns)

CQ	Calling any station	R	all received OK
$\overline{AR}$	end of message	AS	please stand by (AS 5 Wait 5 Minutes)
K	invite any station to transmit	$\overline{SK}$	end of contact (sent before call)
KN	invite a specific station to transmit	CL	going off the air (closing down)
$\overline{BK}$	invite receiving station to transmit	$\overline{BT}$	(also shown as =) break between paragraphs

Fig 2.2: Morse characters and abbreviations

er in learning the whole code at once. Some people advocate learning small groups of three or four letters, but I feel that learning the complete alphabet is the way to go. **Fig 2.2** shows a definitive list of characters, punctuation and prosigns that should be learned. To these should be added some of the Q-code, and abbreviations used a lot on CW (**Figs 2.3 and 2.4**).

You must remember that dots are dits, and dashes are dahs. Looking at the letter A, for example, you might think it is dot-dash. This is the wrong approach. It is, in fact, dit-dah, pronounced di-dah. The 't' remains silent. Learning all the characters in this fashion will give you a fundamental sound in your head. Carrying it on to the letter B, it would be dah-di-di-dit. The last dit has the 't' on it this time. I am sure you can follow this, so take the whole alphabet in this fashion and try to get help from a family member. Get them to ask you letters and numbers at random, with you converting them to Morse in response until there is no thinking time at all. The code in di-dah fashion is shown in **Fig 2.5** Morse Characters in di-dah fashion and learning the code in this way makes it sound like Morse and will also make it easier for the student to repeat the code out loud.

The first step to learning Morse is to study the alphabet and numbers and commit them to memory so that you can instantly tell someone the code for any letter or number. Then add some punctuation, Fig 2.2 gives you the list. There are only four commonly used punctuation marks, comma, full-stop, question-mark and slash-bar. Once you have done this, it is time to devote time to practice. The old adage applies here, the more you practice the better you get!

Prosigns and Other Procedural Signals

Prosigns are symbols formed by running together two or three characters into one, without the inter-character space, to make an abbreviation for the most common procedural signals. They are usually written with a BAR over the char-

Common abbreviations

AA	All after
AB	All before
ABT	About
ADR	Address
AGN	Again
AM	Amplitude Modulation
ANT	Antenna
BCI	Broadcast Interference
BCL	Broadcast Listener
BK	Break, Break in
BN	All between; Been
BUG	Semi-Automatic key
B4	Before
C	Yes
CFM	Confirm; I confirm
CK	Check
CL	I am closing my station; Call or nickname
CLD	Called
CLG	Calling
CQ	Calling any station
CW	Continuous wave
DLD	Delivered
DLVD	Delivered
DR	Dear
DX	Distance
ES	And
FB	Fine Business, excellent
FM	Frequency Modulation
GA	Go ahead
GM	Good morning
GN	Good night
GND	Ground
GUD	Good
HI	The telegraph laugh; High
HR	Here; Hear
HV	Have
HW	How
LID	A poor operator
MA	Millamperes
MILS	Millamperes
MSG	Message; Prefix to radiogram
N	No
NCS	Net Control Station
ND	Nothing Doing
NIL	Nothing; I have nothing for you
NM	No more
NR	Number
NW	Now; I resume transmission
OB	Old boy
OC	Old chap
OM	Old man
OP	Operator
OPR	Operator
OT	Old timer; Old top
PBL	Preamble
PSE	Please
PWR	Power
PX	Press
R	Received as transmitted; Are
RCD	Received
RCVR	Receiver
RX	Receiver
REF	Refer to; Referring to; Reference
RFI	Radio frequency interference
RIG	Station equipment
RTTY	Radio teletype
SASE	Self-addressed, stamped envelope
SED	Said
SIG	Signature; Signal
SINE	Operator's personal initials
SKED	Schedule
SRI	Sorry
SSB	Single Side Band
SVC	Service; Prefix to service message
T	Zero
TFC	Traffic
TMW	Tomorrow
TKS	Thanks
TNX	Thanks
TT	That
TU	Thank you
TVI	Television interference
TX	Transmitter
TXT	Text
UR	Your; You're
URS	Yours
VFO	Variable Frequency Oscillator
VY	Very
WA	Word after
WB	Word before
WD	Word
WDS	Words
WKD	Worked
WKG	Working
WL	Well; Will
WUD	Would
WX	Weather
XCVR	Transceiver
XMTR	Transmitter
XTAL	Crystal
XYL	Wife
YL	Young lady
73	Best Regards
88	Hugs and Kisses

Fig 2.3: Common abbreviations used on Morse code

Q Signals (they take the form of a question only when followed by a question mark)

Code	Question	Answer
QRG	Will you tell me my exact freq (or that of ___)?	Your exact frequency (or that of ___) is ___ kHz.
QRH	Does my frequency vary?	Your frequency varies.
QRI	How is the tone of my transmission?	The tone of your transmission is ___. (1 Good, 2 Variable, 3. Bad)
QRJ	Are you receiving me badly?	I can not receive you. Your signals are too weak.
QRK	What is the intelligibility of my signals (or those of ___)?	The intelligibility of your signals (or those of ___) is: (1 Bad, 2 Poor, 3 Fair, 4 Good, 5 Excellent)
QRL	Are you busy?	I am busy (or I am busy with ___). Please do not interfere.
QRM	Is my transmission being interfered with?	Your transmission is being interfered with ___. (1 Nil, 2 Slightly, 3 Moderately, 4 Severely, 5 Extremely)
QRN	Are you troubled by static?	I am troubled by static —. (1-5 as under QRM)
QRO	Shall I increase power?	Increase power.
QRP	Shall I decrease power?	Decrease power.
QRQ	Shall I send faster?	Send faster ___. (WPM)
QRS	Shall I send more slowly?	Send more slowly ___. (WPM)
QRT	Shall I stop sending?	Stop sending.
QRU	Have you anything for me?	I have nothing for you.
QRV	Are you ready?	I am ready.
QRW	Shall I inform ___ that you are calling on ___ kHz?	Please inform ___ that I am calling on ___ kHz.
QRX	When will you call me again?	I will call you again at ___ hours (on ___ kHz).
QRY	What is my turn?	Your turn is numbered ___.
QRZ	Who is calling me?	You are being called by ___ (on ___ kHz).
QSA	What is the strength of my signals (or those of ___)?	The strength of you signals (or those of ___) is ___. (1 Scarcely perceptible, 2 Weak, 3 Fairly good, 4 Good, 5 Very good)
QSB	Are my signals fading?	Your signals are fading.
QSD	Is my keying defective?	Your keying is defective?
QSG	Shall I send ___ messages at a time?	Send ___ messages at a time.
QSK	Can you hear me in between your signals and if so, can I break in on your transmission?	I can hear you between my signals; break in on my transmission.
QSL	Can you acknowledge receipt?	I am acknowledging receipt.
QSM	Shall I repeat the last message I sent you, or some previous message?	Repeat the last message you sent me [or message(s) number(s) ___].
QSN	Did you hear e (or ___) on ___ kHz?	I did hear you (or ___) on ___ kHz.
QSO	Can you communicate with ___ direct or by relay?	I can communicate with ___ direct (or by relay through ___)
QSP	Will you relay to ___?	I will relay to ___.
QST	General call preceding a message addressed to all amateurs & ARRL members. In effect "CQ ARRL".	
QSU	Shall I send or reply on this freq (or on ___ kHz)?	Send a series of Vs on this frequency (or on ___ kHz).
QSW	Will you send on this frequency (or on ___ kHz)?	I am going to send on this frequency (or on ___ kHz).
QSX	Will you listen to ___ on ___ kHz?	I am listening to ___ on ___ kHz.
QSY	Shall I change to to transmission on another freq?	Change transmission to another frequency (or ___ kHz).
QSZ	Shall I send each word or group more than once?	Send each word or group twice (or ___ times).
QTA	Shall I cancel message number ___?	Cancel message number ___.
QTB	Do you agree with my counting of words?	I do not agree with your counting of words. I will repeat the first letter or digit of each word or group.
QTC	How many messages have you to send?	I have ___ messages for you (or for ___).
QTH	What is your location?	My location is ___.
QTR	What is the correct time?	The correct time is ___.

Fig 2.4: International Q-code definitions. Not all are in common use in amateur radio

acters or sometimes within echelons, eg <AR>. In addition to those shown in Fig 2.2, the ARRL lists the following prosigns on their document at http://www.arrl.org/FandES/field/nts-mpg/pdf/MPG304A.pdf:

<AAA>	Full Stop
<KA>	Beginning of message
<VE>	Beginning of transmission (alternative to above)
<AS>	Stand by; wait
<BT>	Separation (break) between address and text; between text and signature
<HH>	(Error in sending. 8 dits - Transmission continues with last word correctly sent.)
<IMI>	Repeat; I say again. (Difficult or unusual words or groups.)
<NR>	Number follows

Letters, number and punctuation

A	di-dah	V	di-di-di-dah
B	dah-di-di-dit	W	di-dah-dah
C	dah-di-dah-dit	X	dah-di-di-dah
D	dah-di-dit	Y	dah-di-dah-dah
E	dit	Z	dah-dah-di-dit
F	di-di-dah-dit		
G	dah-dah-dit	1	di-dah-dah-dah-dah
H	di-di-di-dit	2	di-di-dah-dah-dah
I	di-dit	3	di-di-di-dah-dah
J	di-dah-dah-dah	4	di-di-di-di-dah
K	dah-di-dah	5	di-di-di-di-dit
L	di-dah-di-dit	6	dah-di-di-di-dit
M	dah-dah	7	dah-dah-di-di-dit
N	dah-dit	8	dah-dah-dah-di-dit
O	dah-dah-dah	9	dah-dah-dah-dah-dit
P	di-dah-dah-dit	0	dah-dah-dah-dah-dah
Q	dah-dah-di-dah		
R	di-dah-dit	Question Mark	di-di-dah-dah-di-dit
S	di-di-dit	Full Stop	di-dah-di-dah-di-dah
T	dah	Comma	dah-dah-di-di-dah-dah
U	di-dit-dah		

Procedural Signals

Stroke	dah-di-di-dah-dit
Break Sign (= or $\overline{\mathrm{BT}}$)	dah-di-di-di-dah
End of Message (+ or $\overline{\mathrm{AR}}$)	di-dah-di-dah-dit
End of Work	di-di-di-dah-di-dah
Error	di-di-di-di-di-di-di-dit
Wait ($\overline{\mathrm{AS}}$)	di-dah-di-di-dit
Open Invitation to Transmit (K)	dah-di-dah
Invitation to Named Station ($\overline{\mathrm{KN}}$)	dah-di-dah-dah-dit
Closing Station ($\overline{\mathrm{CL}}$)	dah-di-dah-di-di-dah-di-dit

Fig 2.5: Morse code displayed as it sounds - the best way to remember it

The following without the < > are other commonly used two letter procedural signals. These are usually sent as two separate letters.

BK	Break
CL	Going off the air (clear)
CQ	Calling any amateur radio station (Many add a space between the C and the Q)
DE	This or From
KN	Go only, invite a specific station to transmit

Commonly used in CW - single letter meanings

C	Correct yes
K	Go, invite any station to transmit
N	No Negative
R	All received OK

The conversion of text to Morse can be practised at any time of the day. For example, if you are driving, convert the number plate of the vehicle in front to Morse code. Do the same for advertisements, shop signs and so on. Don't forget to use the "di-dah" method, saying the Morse out loud as you convert.

Tuition Groups

One suggestion I would thoroughly recommend is to ask at your local radio club to find if there is an active Morse tuition group. This would be well worthwhile joining, not only for the one-to-one tuition that you will receive, but competition within the group will also help your progress. Not only that, it is a lot of fun too, and it relieves any stress that some might have. I have taught Morse for a number of years now and I find that group learning is the best way to get the best from each member of the group. A light-hearted approach also helps, whilst maintaining steady progress. Normally it takes about 12 weeks to reach 12WPM with one meeting per week and roughly 30-60 minutes practice per day.

Having said a light-hearted approach is fine, progress is dependent absolutely on the amount of practice put in, and the group tutor will be able to tell if any has been done, so dedication is mandatory. It is much like learning the piano. Performance is directly proportional to the amount of practice!

Using your Computer

There are several learning programs available as freeware that can assist in several different ways. Links to these, or in some cases the software itself, can be found on the CD provided free with this book.

The excellent AA9PW Morse Practice Program is available on the Internet at: http://www.aa9pw.com/radio/morse.html and is an interactive program. You don't have to download anything, just work from the page that appears on your screen.

I use this program in my Morse tuition classes and find it invaluable. The page you will see when you use the above URL will generate morse code consisting of groups of random characters and there are 25 characters per group (five sets of five characters). You can pick which group of characters you wish to be tested on: the alphabet, the numbers or punctuation (including prosigns) or all three. The computer creates an audio file and sends that to your browser so your browser (or helper application) needs to be capable of handling these files. You will also need a sound card with speakers or similar to be able to hear the Morse. There are several levels, starting with letters and numbers which is the obvious one to start with. You can start as low as 5WPM and increase up to 50WPM if you feel brave enough! The audio files on the CD attached to this book were made using AA9PW's page.

There are many other programs you can use for basic training of course and if you want a varied selection, just search for Morse programs on the Internet (or use the links on the CD) and you will find dozens of them.

Always select a speed just above that which you can receive fully. Once you achieve that level, increase the speed again. In this way, you are striving to achieve complete copy all the time. It is the only way to improve. Once you feel comfortable at around nine words per minute with the letters and numbers, try the punctuation. This will seem easier as there are only four that you really need bother about. When you are happy with those, try some call signs. Mixed letters and numbers always slow you down again, but you should persevere here because obviously call signs are going to be a large part of your use of Morse code on the amateur bands. However, when selecting the speed, use a character speed of 15WPM Farnsworth timing. The Morse will then be sent to you at 15WPM for each character with a large space enabling thinking time. If you start at something like 5WPM, the character speed is too laborious and using the Farnsworth method is by far the best way of learning. This is discussed in Chapter 3.

Don't forget that the learning process takes time and is proportional to the time that you devote to it. There is nothing magic in this process, just a lot of hard work.

3

Morse at 12-15WPM

THERE ARE SEVERAL methods of learning Morse. Some are quite abysmal, suggesting learning opposites, or visualising Morse code as heaped numbers of dots and dashes, or worse still, sending it on a key. Do not touch a Morse key until you have been receiving Morse for some while and are receiving competently around 12 to 15WPM. If you do, you will develop bad habits which will be difficult to cure. From teaching the code for years and having quite a good success rate, I prefer using the Farnsworth method.

The Farnsworth timing came from the late Donald R "Russ" Farnsworth, W6TTB, who in the late 1950s asked Bart Bartlett, W6OWP, to help him prepare some tapes for a code course he had developed. Farnsworth's unique method of instruction was to maintain the code speed for individual characters but allowing longer spaces.

Farnsworth timing is defined as sending the characters at a faster speed than the words. For example, sending the characters at 20WPM but adding enough time between them to slow down the overall rate to 10WPM gives enough thinking time between characters while maintaining the actual character rate at 20WPM.

Once you reach 20WPM, the spacing is correct for that speed and you can then progress further either by doing the same again or leaving the spacing correct for the speed, allowing true timing. Having reached 20WPM, it is experience and use of the mode that will improve your speed.

Standard Timing

Standard timing is as follows:

The period of a single dot is one unit, measured in seconds.

A dash is a period of three units.

A period of one unit separates each element (dot or dash) within a character.

A period of three units separates each character within a word.

A period of seven units separates each word.

A graphical representation of the timing is shown in **Fig 3.1**

The problem in learning Morse is that at speeds above a few words per minute, Morse is most easily read by ear when the characters are recognized rhythmically rather than by remembering the dots and dashes. But the person just learning Morse starts at very slow speeds, where thinking about the actual dots and dashes is easier than recognizing the slow rhythm of the characters. So in order to increase their ability to read Morse above a few words per minute,

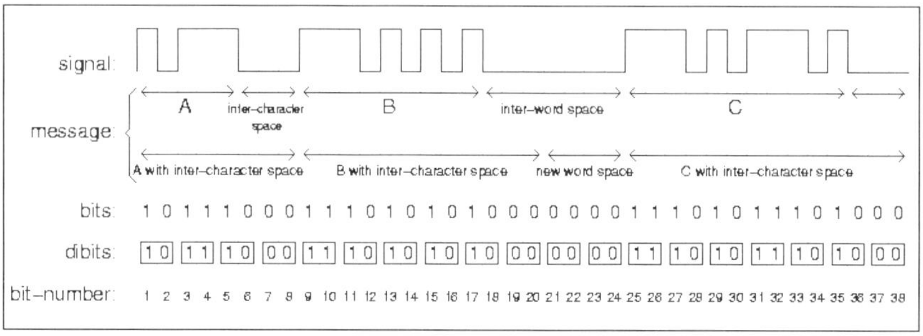

Fig 3.1: Morse timing shown graphically

students are forced to shift from the 'thinking of dots and dashes' mode to the rhythm recognition mode. Along the way, there will be barriers which at the time seem insurmountable. This is the reason why many just give up. Donating a few minutes per week is not going to crack it! Dedication is needed, and I mean dedication. I usually quote the plate of bacon and eggs. The chicken "donated" to the breakfast plate, but the pig was dedicated! (**Fig3.2**)

The idea behind the Farnsworth method is to eliminate the counting phase by sending the characters at a speed at which rhythm recognition is easy and counting is not. This forces the student to learn the rhythms. Initially, the rate of transmission is slowed by the addition of time (between characters) to allow the student gradually to build the skill at recognizing and writing the received text. The process by which the student recognizes the characters is never changed; he just gets better (faster) at doing so.

Recently, the ARRL finished converting all of its Morse material to Farnsworth timing. They are using a standard of sending transmission at an 18WPM character rate, but of course, at 18WPM and faster speeds, the ARRL transmission reverts to standard timing, since no extra time has to be inserted. This standard applies to all code practice and test tapes, and to W1AW transmissions.

Morse Training by the Koch Method

Koch's method is a simple, direct way of building reflexes. However, it requires either a computer and Morse software or a personal trainer. That's why it was overlooked for so many years. Now that computers are commonplace, it could become the standard Morse training method. Here's how it works:

You start out by setting up your computer to send you Morse characters at 20WPM and at an overall sending speed of at least 15WPM. You then get out your paper and pencil and have the machine start sending - but only two characters. That's right, for your first sessions, you'll only have two choices. Copy on paper for five minutes, then stop the machine and compare what you copied with what the machine sent. Count characters and calculate your percentage of correct copy.

Fig 3.2: Pig was dedicated, chicken not quite so

If your score is 90 percent or better -- congratulations! You just learned your first two characters, and, importantly, you learned them at full speed. You'll never have to learn them over again. If you failed to achieve a ninety percent success rate, do some more practice. As soon as you can copy the first two characters with 90% accuracy, add a third character to your practice. Your accuracy will drop as you work on assimilating the new character, but it will rise again to 90% or better. Then you add the fourth character, and so on.

This method does not allow you to build that lookup table in your brain. To copy at full speed, you **must** build the reflexes in order to achieve 90% accuracy. And that's what you're spending your time doing -- building reflexes. Think of it as a parallel to perfecting a tennis stroke or mastering a gymnastic routine; you're practicing until you get it right. The Koch method of building code proficiency character-by-character is similar to standard methods of teaching touch typing, another skill that must be reflexive.

This is a very individual method of training -- you progress at your own best speed, and spend only the time required to gain each new character. This means that you will waste no time in reaching your goal.

How much time is required? That will depend on the individual. You can get an idea of how long it's going to take after you've mastered a few characters. Keep track of your training sessions (some software will do this for you) and calculate your hours-per-character rate (or characters-per-hour if you're really fast!). That, multiplied by the amount of characters will give a rough idea of how long it's going to take.

While the Koch method is the fastest method of Morse training, speed alone is not its principal advantage. Its principal advantage, and a major difference from other methods, is that it provides you with constant positive reinforcement. This begins with your realization, after mastering the first two characters, that you **can** copy code at 15 or 20WPM, because you just did it. After that, each new character mastered is further proof of your progress. Contrast that to slowly trying to build speed up from 4 or 5WPM, then hitting the plateau at 10WPM and seeing no progress for a long time. With the Koch method, frustration is at a minimum.

Constant testing is necessary to ensure that you maximize the effectiveness of the Koch method. You must copy on paper, so you can grade yourself. Remember, if you score 90% accuracy or better, add another character. If you score any less than that, try again. By constantly testing yourself on continuous copying of at least five minutes, you know exactly how you're doing and exactly when you should add another character. This results in the fastest progress possible. This is a good method of self-teaching and combined with the Farnsworth method with a tutor would be quite productive. A Koch trainer can be dowloaded from http://www.g4fon.net/.

The Beauty of Morse

For someone with no knowledge of Morse Code at all tuning across the CW end of an amateur band such as 20 meters, it will sound like a lot of noise, heterodynes, and interference. For someone with ability to use Morse at around 25-30 WPM tuning across the same band can produce a world of its own. With a pair of headphones on, a lively band, and a little time, the skilled operator can find new countries, listen to conversations, find the ARRL W1AW News broadcast, find old friends and stop for a chat, or if there happens to be a contest running, he can hone his operating skills even further. Some of the Russian CW operators use high-speed Morse, and speeds of 40-50WPM are not uncommon.

You may feel like a change from the AA9PW code that your computer generates. It is a good idea to have a change too. A different sounding note, different content and style, some plain language perhaps; in that case try the ZL1AN Morse Code Teaching Program which can be found on the CD provided with this book.

From the start screen **(Fig 3.3)** you can select from 950 pieces of text. A lot of them have numbers in them too, so it provides the student with a varied selection of characters. The parameters can be set to please the student too, including the pitch of the note, speed, and so on. This program can provide quite a number of hours of practice, but can get quite boring if you don't like cricket, as a lot of the text is concerned with cricket reports!

Once a speed of around 15WPM has been attained, don't leave it there. Carry on with the regular practice and set yourself targets, 20, 25, 30WPM, and so on. Code proficiency certificates are still available, but remember, to obtain these it will be necessary to write down what you are copying. This is not easy once you get over 25 wpm unless regular practice is maintained. As you attain a higher speed, listening becomes more of a pleasure and not a chore, so try to improve all the time. Also, remember that Morse is an art form, as well as a skill. Take a pride in your ability at both receiving and sending as near perfect Morse as possible. If you send good Morse, you will never be short of a contact. It is so much easier to read well sent Morse and a pleasure to listen to.

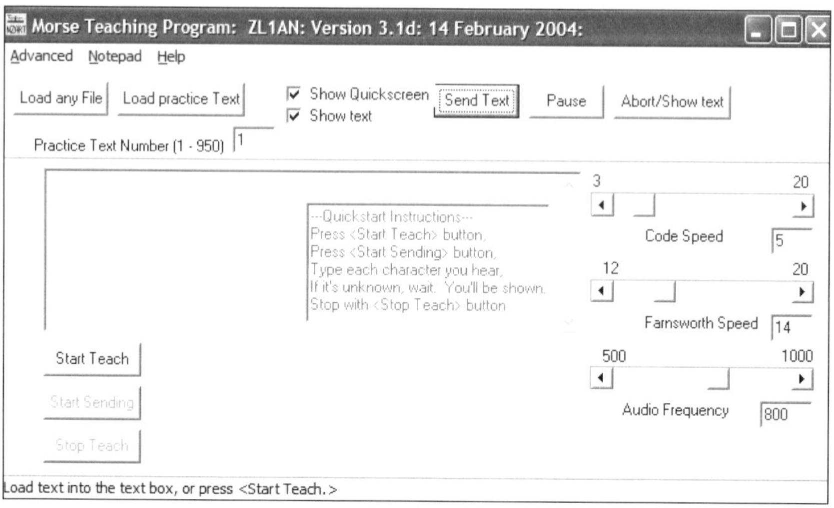

Fig 3.3: Screen showing ZL1AN program after downloading

4

Keys and Sending

Fig 4.1: The author's straight key was bought decades ago and is still going strong

BY THIS TIME, you will have been dedicated enough with your practice and have now mastered around 12 to 15 words per minute fairly comfortably. Morse is now beginning to feel like an exciting mode, and you copy without really thinking about the translation from dots and dashes to letters and numbers. At this stage, it begins to be an automatic translation within the brain and also like another language, albeit somewhat slow. However, the character formation is embedded in your brain too, hopefully with the correct mark to space ratio and spacing, so now is a good time to turn to sending.

Straight Keys

A straight key is the common telegraph key seen in old movies. It is a simple bar with a knob on top and a contact underneath. When the bar is depressed against spring tension, it forms a circuit and allows electricity to flow. Traditionally, American telegraph keys had flat topped knobs and narrow bars (frequently curved). British telegraph keys had ball shaped knobs and thick bars. This appears to be purely a matter of culture and training, but the users of each are tremendously partisan.

The straight key is simple and reliable, but the rapid pumping action needed to send a string of dots (or dits as most operators call them) poses some significant drawbacks. Transmission speeds are limited to around 20 words per minute, and in the early days of telegraphy a number of professional telegraphers developed a repetitive stress injury known as glass arm or telegrapher's paralysis.

Using a Morse Key Under Instruction

Take advice from an experienced CW operator when buying a straight key. There are many variations available, some very poor, others mediocre, and some very good. It should be roughly 100mm by 180mm and weigh around 600 to 900 grams. It should be on a heavy base and made from brass, with a good centre pivot. It must have a space adjustment and good clean keying contacts, preferably silver plated. On some keys it is possible to adjust the tension as well. It should also have a standard 'doorknob', or mushroom with skirt, type knob, to allow proper holding of the key.

Prices vary, and will not be cheap. However, bear in mind that a straight key will last a lifetime and will always have a place in the shack of a true amateur. You can pay anything from around £60 up to a huge £600 for a really super-duper chrome plated key on a hand-polished serpentine base. The key I have had for decades is still going strong and is shown in **Fig 4.1**.

Sending on a Straight Key

The purpose of **Fig 4.2** is to demonstrate the correct way to hold a Morse key. It is quite amusing to see how the whole Morse procedure is abused in some films. I have even seen a Morse key keyed with just the index finger, not that any real Morse was being sent!

The first thing to do is to sit square to the table, with the key in line with your keying arm. I say this, because you may be left handed and prefer to key with the left hand rather than the right. Have the Morse key toward the front of the table, so that it is not possible to rest your arm on the table. The Americans teach sending on a straight key with the forearm resting on the table, but I think this leads to sloppy Morse. Hold the key with the index finger on the knob, and the thumb and middle finger on the skirt, as shown in Fig 4.2.

I would prefer your instructor to show you how to key properly, as it does take some practice to get it correct. If you hold the key as shown in the picture, you should then key from the wrist. I call it the 'limp wrist' method, using the arm

Fig 4.2: How to hold the Morse key

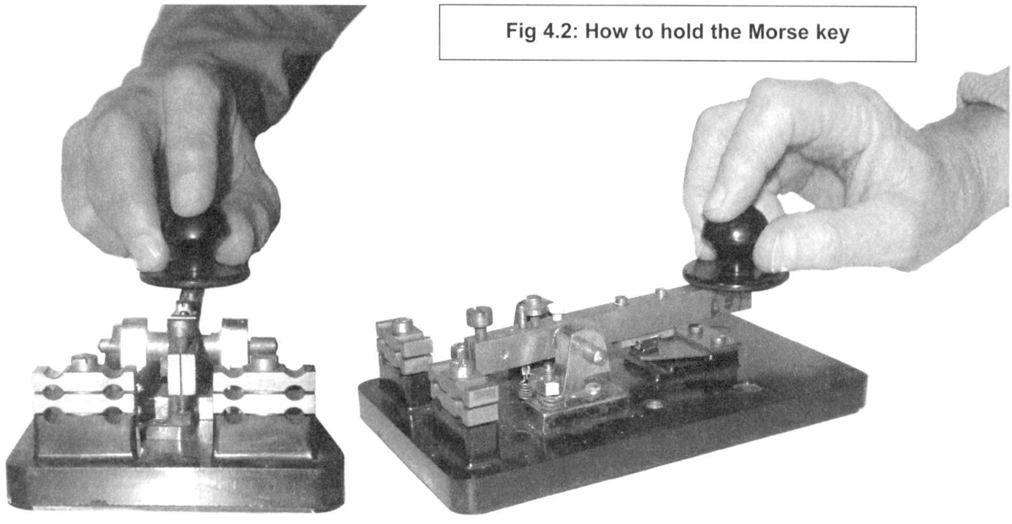

as a lever, with the wrist being the actual point at which you form the Morse characters. Holding the key will then make the key follow your wrist. It is not easy to explain, but must be carried out correctly, hence my emphasis on an instructor. Bad habits are extremely difficult to break, and lead to bad Morse, which will inevitably lead to zero contacts!

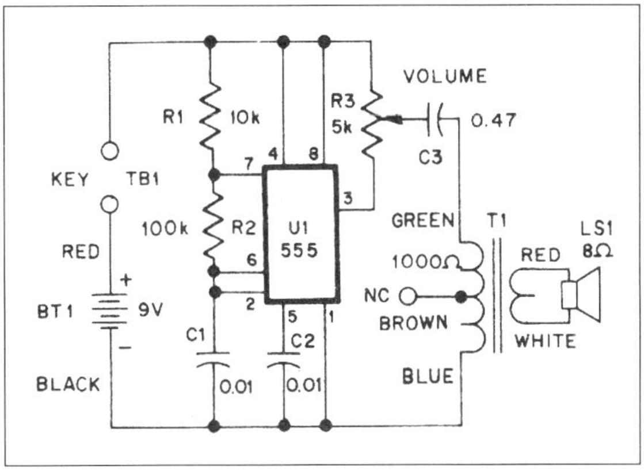

Fig 4.3: Morse practice oscillator

Practice Makes Perfect

Your first sending session will preferably be with your instructor making sure that all is well with the technique. You will need to have a Morse practice oscillator. You can make one fairly easily on a piece of Vero board, using a 555 timer chip. A suitable circuit is shown in **Fig 4.3**.

This takes a while to master in itself, but will bode well for the future, so time spent in doing it properly is time well spent. Practice at home, as you did with reception, for around 20 minutes a night minimum and try to emulate what you hear in the Morse program. A few weeks of this, gradually increasing the speed will ensure that your Morse will be as perfect as can be, providing you actually do the practice. Once your instructor is happy with your results at around 18WPM, receiving and sending on the straight key, then it is time to think about a paddle.

QSO Mode

Before discussing paddles, it might be a good idea to see the format of a basic Morse contact as it might happen on the air. There are many ways of making a contact (or 'QSO'), the standard rubber stamp short QSO (used mainly with those that have little command of the English language), contest QSO, DX style, and 'rag-chew'. The latter, if both stations let it, can carry on for an hour or more, depending on what you have to say to each other. Here are some examples:

Example 1:

CQ CQ CQ de G9XXX G9XXX G9XXX AR PSE K

G9XXX de G9YYY G9YYY AR KN

G9YYY DE G9XXX GM = TU fer call = Name Chris QTH Norwich = UR RST 599 = Hw? G9YYY de G9XXX AR KN

G9XXX DE G9YYY TU Chris GM = UR RST 599 = Name Bernie QTH LONDON - Nice meet u fer 1st time = Ant 3 el Yagi up 30 m = Pwr 100 W = Wx hr fb temp 30 C = G9XXX de G9YYY AR KN

G9YYY de G9XXX Name agn? BK

BK Bernie Bernie BK

BK Thanks Bernie = Sri had QRM = Hr ANT is 2 el up 22 m es 1TT W = Wish I had ur 3 el = hr no space fer big `ants = Wx hr also fine temp 26 C = hv bn ham 1 yr, still getting used to cw = condx rotten past few days = hv hrd vy few stns on 40 m = only sum W6 on LP b4 sun-set = hvnt had much luck wrking them tho = ok must run hv to do sum chores in the house = 73 cul bernie AR G9YYY de G9XXX AR SK CL`

`G9XXX de G9YYY OK Chris = Yes 3 el lot of fun = Wrk W6 on LP all the time = Condx rotten hr too but wrked FO stn on SP at sunset = OK 73 tnx QSO es QSL via buro = G9XXX de G9YYY SK`

Notice how the term "BK" was used to quickly pass the transmission back to the other station. This short QSO was reasonably chatty, but it assumes knowledge of the abbreviations, plus also some knowledge of DX calls and propagation. Also, note that Chris gave his power as 1TT W, using "cut" numbers, a T for a 0. This is quite common.

Example 2 (a directional CQ):

`CQ OC de G9BBB CQ OC de G9BBB G9BBB OC KN`

`de IK1XYZ`

`CQ OC de G9BBB CQ OC de G9BBB G9BBB OC KN`

`de IK1XYZ`

`IK1XYZ pse qsy ur not in OC = CQ OC de G9BBB CQ OC de G9BBB G9BBB OC KN`

Note the KN at the end of a directional call to Oceania. The IK is just trying it on, he knew that G9BBB was calling OC only, but tried anyway, not very good operating. He should have respected that directional call.

Example 3 (a complete contest-style QSO):

`CQ TEST G9ZZZ G9ZZZ`

`G9VVV`

`G9VVV 5NN 13N`

`5NN TT1`

`TU G9ZZZ TEST`

In contests, ending signals, the "DE" and other niceties are generally dispensed with. Notice how the example QSO includes no ending signals, and no instances of the "DE" signal. All contesters are aware of what is going on. Note the cut numbers again.

Example 4 (planning for your first QSO):

For your first few contacts, write down everything you are going to send, in advance. When the other stations sends, just write down what he/she is sending. Keep the abbreviation list handy. Don't get too adventurous initially; just stick to the standard script until you find your feet. Here is a suggested first QSO:

`QRL?`

`CQ CQ CQ  de G9RRR G9RRR AR PSE K`

`G9RRR DE G9PPP G9PPP AR PSE KN`

`G9PPP de G9RRR GM = Name Bill QTH HULL = UR RST 599 = This is my 1st QSO on CW HI = G9PPP de G9RRR AR KN`

`G9RRR de G9PPP OK Bill = all OK = UR RST 579 579 = Name is Sid Sid es running 100 W to a dipole = Hw? = G9RRR de G9PPP AR KN`

After the initial couple of overs, you can either finish if you are nervous, or carry on. If the other station is a good operator he will slow down to your speed level and also take into account that it is your first CW QSO.

Alternative Key Designs

The first widely accepted alternative key was the sideswiper or sidewinder, sometimes called a cootie key. This key uses a side-to-side action with contacts in both directions and the arm spring-loaded to return to center. A series of dits could be sent by rocking the arm back and forth. Since few people are able to make the leftward and rightward motions identically, the alternating action produces a distinctive rhythm or swing. Experienced telegraphers can recognize one another by keying style, nicknamed 'fist', as easily as they can recognize a voice. Although the sideswiper is seldom seen or used today, nearly all advanced keys use some form of side-to-side action.

Undoubtedly the most popular side-to-side mechanical key is the semi-automatic key or bug, sometimes known by as a Vibroplex key, after the company that first manufactured them. When the paddle is pressed to the left it makes a continuous contact suitable for sending dashes (or dahs, as most operators call them). When the paddle is pressed to the right, a horizontal pendulum is set into motion which rocks against the contact points, sending a series of short pulses (dits) at a speed which is controlled by the position of the pendulum weight. A skilled operator can achieve sending speeds of 50 words per minute with a bug.

Like the bug, the electronic keyer operates sideways. When pressed to one side the electronics generate a series of 'dahs' and when pressed the other way, a series of 'dits'. Most electronic keyers include a dit memory function which frees the operator from the need to perfectly time his transitions in the sequence dah-dit-dah. With dit memory, if the operator's keying action is about one dit ahead of the actual transmission, the keyer's output for each letter will be machine-perfect. An iambic keyer sports dual paddles, one for dit and one for dah; pressing both at the same time produces an alternating dit-dah-dit-dah sequence. Electronic keyers allow very high speed transmission of code.

Using an Electronic Keyer and Paddle

Anything over 20WPM will be difficult to maintain on the straight key, and although it is possible to key around 25WPM with the straight key, it is much easier with an electronic keyer and paddle. Most transceivers have electronic keyers built in and you could manage with that and a paddle. However, I prefer an external keyer, for two reasons. One is that it can be used to key other transceivers with the same memory functions, plus all the other options that external keyers have. The other is that it has more versatility in some situations, such as contesting with a computer, and practising off the air.

I have an AEA MM3 Morse Machine **(Fig 4.4)**, and have used that for around 15 years now. Prior to that I had a home-made TTL memory keyer. The MM3 has a super side-tone, something that is very important to a CW operator, and also lots of pre-programmable mem-

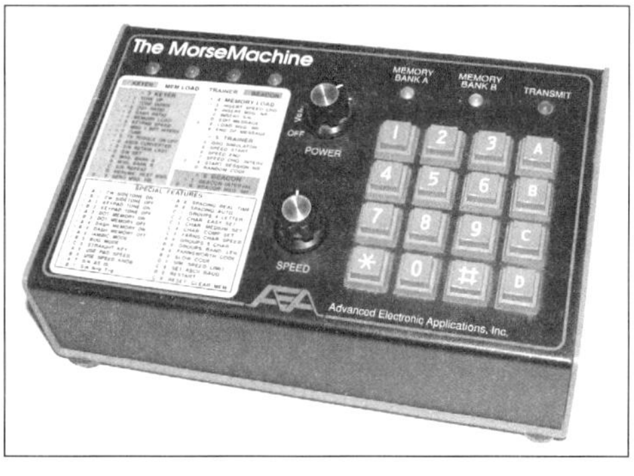

Fig 4.4: AEA MM-3 keyer

ories that have various uses. It also has a Doctor DX feature, a QSO and pile-up contest feature, and these can be very useful for off-air practice and tuition. The paddle I use is a Vibroplex single lever paddle (**Fig 4.5**) that I bought around 40 years ago, and it is still going strong! The latest models are very nice indeed, and are a superb addition to the shack, but the prices are a bit daunting however.

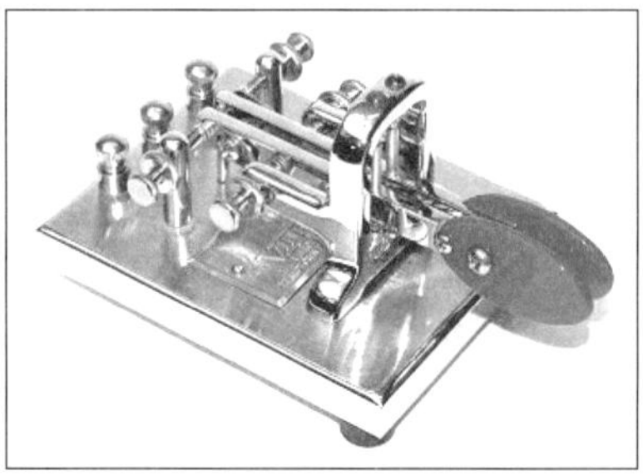

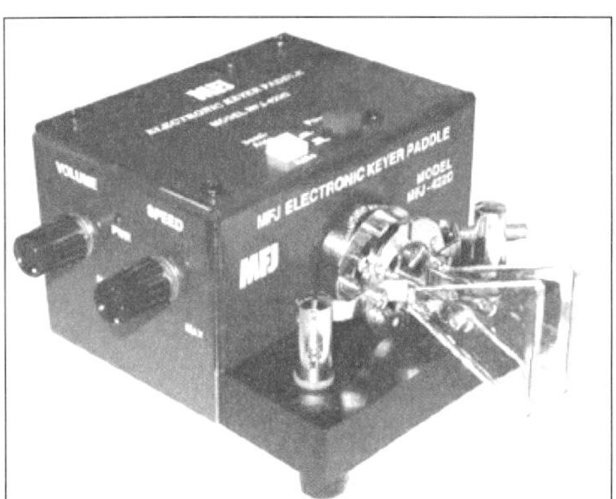

(Top to bottom:)
Fig 4.5: Old Vibroplex Paddle
Fig 4.6: Modern Vibroplex Iambic paddle
Fig 4.7: Paddle made by Bencher
Fig 4.8: Combined keyer and paddle from MFJ

Have a look at paddles and try several before you decide what to buy. The Iambic might be a better one to have, bearing in mind that your key will probably last for decades. The Vibroplex Iambic paddle is shown in **Fig 4.6**.

Changing to this method will again require lots of practice before you use it on the air, but it will allow you to build up your speed to 40WPM plus if you practice enough. Using a paddle is the best way of using CW on the air with maximum comfort and minimum effort and is used by all good CW operators. Of course, you could take it even further and throw away the key altogether and use a computer keyboard if you can touch type! I regard this as cheating for everyday use, but it is extensively used in DXpeditions and contests.

There are other makes to choose from. Vibroplex have been in the business the longest, hence the fact that some amateurs call their paddle a 'Vibroplex'. However, there are some very good keys made by other manufacturers. For instance, MFJ make a selection, as do Bencher (**Figs 4.7 and 4.8**).

Some paddles can be made on special order. This costs a lot of money but you will have an individual key with the characteristics that you want all in one key, whereas you sometimes find that you have to compromise.

There are even designs to be found on the Internet for those that can use a lathe and make their own. Several amateurs have done this and made a first-class paddle.

Do take a while looking and trying several types and makes before buying. The paddle you buy will have a large influence on your sending and it is essential that you feel comfortable with it.

5

Progress at Last

BY THIS TIME, you will have devoted enough time and patience to become proficient at both receiving and sending with an electronic keyer at around 15-20WPM. Listening on the HF bands has revealed that conversations are taking place, not merely an exchange of dots and dashes, and you are enthusiastic enough to want to work some DX. Obviously you have mastered all the abbreviations too. You will have had lots of contacts on the air at a speed you can manage comfortably, in order to gain experience in doing just that. When you feel completely at ease with this, it is time to take the next step.

In this chapter you will have a whole new learning curve if you want to be a really good operator. Just as you thought the practice was all over, here's news for you - it isn't!

Keyer Construction

If you are enthusiastic about making your own electronic keyer, there are a number of circuits around. The RSGB has published several in recent years, plus you can find them in various publications and on the Internet.

This circuit is by N1HFX and is quite straightforward to build for a reasonable amount of money. It uses only two inexpensive ICs which are available at most stores such as Maplin or RS. It lacks 'bells and whistles', but is capable of producing perfectly sounding CW. The total cost should not exceed £15 for the parts. It is best to construct it in a metal enclosure to avoid any RF interference.

The circuit (**Fig 5.1**) uses a quad NOR gate chip and a dual D flip-flop chip. Two NOR gates are used as the clock generator and the frequency is determined by C1, R1 and R2 which sets the keying speed. This clock signal is fed into Pin 3 of IC 2 which is the dot flip flop.

When the dot paddle is depressed (grounded), pin 1 of the flip flop changes states for as long as the dot paddle is depressed. When the dash paddle is grounded, this in turn causes the dot paddle to be grounded also through Diode D1. This starts the dash cycle which continues until Pin 13 of IC-2 changes state again. Diode D2 keeps the dot paddle low until the dash cycle finishes. Pins 1 and 13 of each flip flop form the output to transistor Q1. Because Pin 1 is not always high during the dash cycle, Diodes D3 and D4 form an OR gate to keep Q1 saturated during a dash cycle.

This circuit has one anomaly which operators should be aware of. The dash paddle always has priority and attempts to squeeze the paddle (both dot and dash depressed) results in a continuous stream of dashes. This means the operator must release the dash paddle before depressing the dot paddle. Experienced

Fig 5.1: Circuit of N1HFX keyer. Many more simple projects by Mike Martell, N1HFX, can be found on the web at www.rason.org

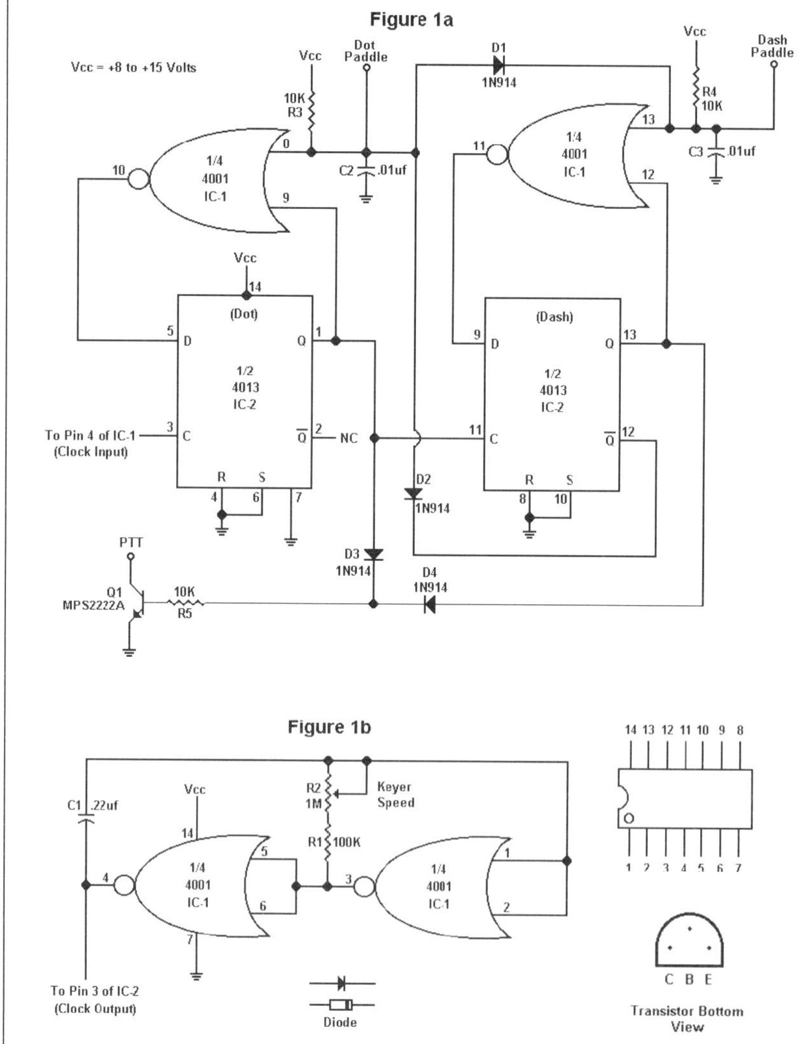

R1	100k	C1	0.22µF (224)	Q1	2N222A or similar NPN
R2	1M	C2, 3	0.01µF (103)	IC-1	4001 Quad NOR gate
R3 - 5	10k	D1 - 4	1N914 or similar	IC-2	4013 Dual D flip-flop

CW operators will have no problem with this circuit but sloppy high speed operators will need to refine their sending since it is not as forgiving as some keyers.

This circuit uses CMOS chips which require special handling to prevent static charges. Sockets are highly recommended. The circuit requires very low power and can be powered from 8 to 15 volts. A nine volt battery should last over a year, meaning an on/off switch isn't needed. No ‘weighting’ control is provided because experience has shown that it is often misused by operators causing code which is difficult to copy.

Adjust the paddle to a very light action. You don't want to be slapping the key all about the table top! Good paddle keys are heavy for this reason. If the paddle is sliding on the desk, you can now buy a sticky-pad from most car accessory shops for about £5. It is a piece of plastic that retains its sticky properties. It will stick to the desk like glue, and the key will stick to it. It is completely removable and washable, because after a time it attracts dust and loses some of the stickiness. With washing, this is soon restored.

Now begin practice to send *quickly*. This sending practice also works wonders on the way to becoming a QRQ operator. In doing this you must form words to express ideas in your mind, while simultaneously sending the thoughts out as CW. This inverts what has been going on in your mental processes to receive. As you increase your speed ability, you will not even be thinking 'letter to Morse' translation, but will be mentally and automatically sending as if it were another language with which you have become quite comfortable.

You will be training your mind on CW in such a way that when it can send fast, it will use the same subconscious patterns to receive fast. There is no short cut method to this; you just have to sit there for a few hours practicing. Sorry if I keep repeating this, but really there is no substitute.

Try to arrange a few skeds with friends at the same level, and spend a few hours a week on the key just chatting. Eventually this will seem just as good as normal voice communications. Gradually increase your speed and keep the competition going, this is a good way to get through the 30WPM and even 40WPM barriers. After that it becomes a problem sending, even with a paddle! Some amateurs have abandoned the paddle in preference to a keyboard.

Above all, do try to have a few CW contacts each day. Don't rest on your laurels, thinking that once you have attained a good speed that it will remain with you. Even a top notch concert pianist still practices for several hours each day. Pick a few contests to try; just taking part will hone your skills very well, and you will eventually find that using a keyer and log book really is slow compared to those using a computer. Time to change again.

Operating Contests

There is more to encourage you to practice. In order to become a proficient contest operator, you now have to learn how to touch type on a computer keyboard. Hopefully, especially in these days when computers are common place, you will have this skill already. All serious contesters use a computer program now; it would not be possible to be competitive without. Several decades ago, logging itself was a skill, assisting the operator to maintain a good QSO rate, but it is far too slow by modern standards.

Of course, there's nothing to prevent you using a key - even a straight one - in a contest if you simply want to improve your technique and speed, rather than compete with the 'big guns'.

Contesters use computer logging programs which do most of the sending for them, as well as the log tasks and checking for duplicates; some programs do even more! But to use them efficiently to contest, you have to be a pretty good keyboard operator as well, and also know the use of the function keys, without using crib notes or key overlay guides during the contest. By learning to touch-type, rather than using just two fingers, your typing skills will be able to keep up with, and surpass, your Morse skills. This will be an advantage when running a contest or pile-up enabling you to drink your tea or eat a sandwich at the same time.

Computer Programs for Morse

These are many and varied. They range from the very basic to ones that have a *wow* factor, enabling you to simulate a contest divorced from the radio, but with all the reality of using one. You can use training programs to keep up your receiving skills. I teach Morse and I use three programs here. You have lots to choose from however. The three I use are as follows.

AA9PW (www.aa9pw.com/radio/morse.html): I find it very useful for basic practice. This has groups of letters, numbers, mixed characters, punctuation, pro-signs, call signs and a mixture of all the above. It also has a news feed that converts to Morse.

Morse Teaching Program by ZL1AN: For a different tone and general practice. It can be found on the AC6V Morse Page, along with many others. This one has a very varied selection of pieces of text, with mixed numbers and letters. Both this program and the one above utilise the Farnsworth method.

MorseRunner (www.dxatlas.com/MorseRunner/): For advanced practice. This program is ideal for using off air as a contest simulator. Written by Alex Shovkoplyas, VE3NEA, who also wrote DXAtlas, it is a look-alike for N1MM, with a very similar layout. Speed, pitch and receiver bandwidth are all adjustable, and also programmable are QRM, QRN, QSB, and flutter, together with the amount of stations calling at any one time. It is very realistic, is superb for handling pile-ups and it can be very addictive! You can download an audio file of someone running a pile-up and there is a league table of those having the highest scores.

The above software and links to a great deal more can be found on the CD provided with this book.

Computer Programs for Contesting

Try to visit a local amateur running a modern contest program, such as *Writelog* (**Fig 5.2**), *N1MM* (**Fig 5.3**), *EI5DI*, *Winlog*, etc. Have a look at all of them.

I have used *WriteLog* in all of the RTTY contests that I have entered, along with all the RSGB Cumulative CW contests. It behaves well, but you have to pay periodically to maintain the latest version. *N1MM*, on the other hand, is freeware and will do a very similar job to *WriteLog*.

Installation of the contest programs is not easy, and requires a fair amount of time. However, there is a lot of useful information provided in the form of help files and a (PDF) handbook with each. Assuming you are fairly computer literate, it should be straightforward. Also, there is probably someone local to you already running one of them.

Using a contest program does require some practice. Now, where have you heard that before! It is another skill that you will need to spend time acquiring. Using such a program on CW still needs the operator to read the Morse code and enter the call into the log window, together with the received serial number and report, but the computer does the sending and increments your serial number. It is good practice to have the paddle and keyer paralleled with the PC keying cir-

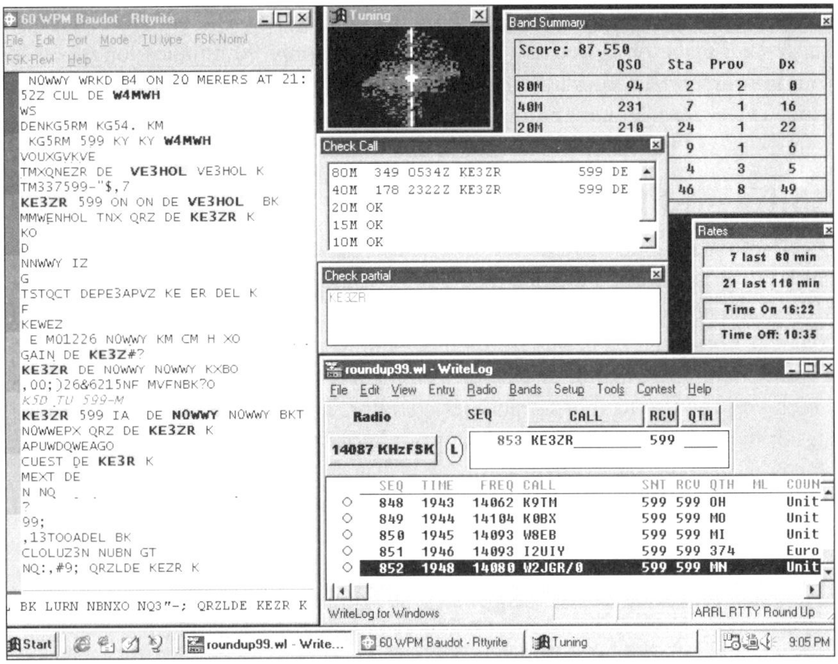

Fig 5.2: The *WriteLog* screen used in a RTTY contest

cuit for the odd comment not covered by the program. All the standard responses are programmed into macros within the program itself, but it is useful to be able to interject a few personal comments during the contest.

Our local club uses *SD*, the contest program written by EI5DI. This is a simple CW program that doesn't have the bells and whistles of the likes of *N1MM* and *WriteLog*. However, it has worked on HF Field Day for us for a number of years.

NOTE: A copy of EI5DI's *SD* program and links to *N1MM* and *WriteLog* can be found on the CD provided with this book.

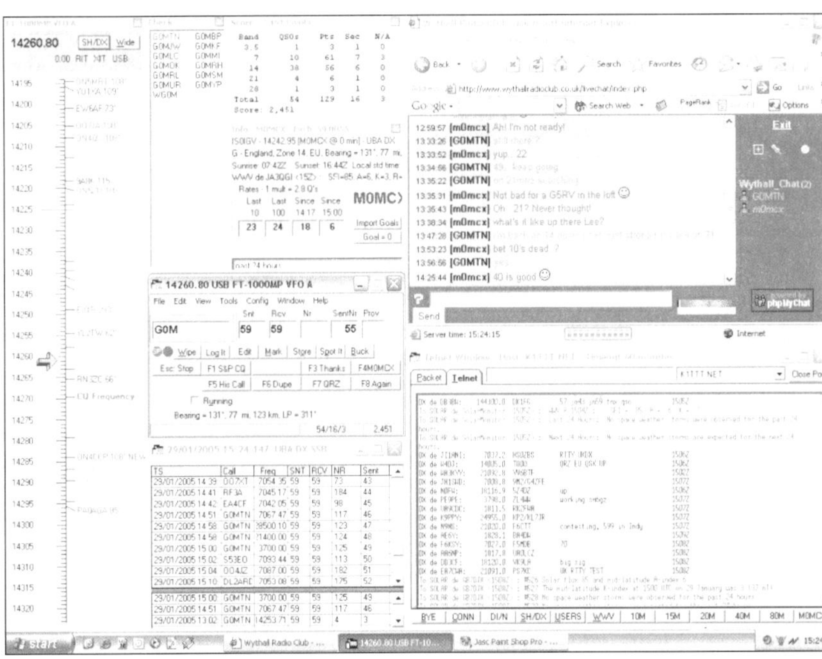

Fig 5.3: The *N1MM* contesting program

A good place to start CW contesting would be the RSGB Cumulative events. These take place on a weekday evening, and are only 90 minutes long. If it is your first time in a contest, there is a slow section at the top end of the CW band which caters for beginners and inexperienced. Good operators will slow down to your speed and have a bit of patience in order to encourage more use of the mode.

Clubs and Newsletters

If you operate a lot of CW on the HF bands, you may receive a very nice unsolicited surprise through the post one day. If you are a good operator, send good Morse, and are willing to talk to others on Morse, rather than have rubber-stamp contacts, you could be in line for membership to the A-1 Ops Club. You receive a membership certificate, like an award, membership details, with some cards which you can use to sponsor another operator. This is all done without soliciting for membership and without your prior knowledge so it adds to the pleasure when you receive your certificate.

You can also join the FOC, if you are lucky! This is the First Class Operator's Club and they have a limited membership, so you might have to wait a while!

FISTS is the International Morse Preservation Society. They have a regular newsletter, operating calling frequencies, activities, printed log-books and so on.

There are others, so as you see Morse is far from dead, and in fact is unlikely to die within amateur radio.

Fast (QRQ) Receiving Practice

While waiting to acquire and set up some of the practice programs, do on-the-air receiving practice. Find QSOs in progress at a speed you can comfortably copy down with a pen. Now, put the pen down, and just start listening. Try to recognize the letters as they come along, and group them into words on your mind's 'blackboard'. It's not easy to do, but keep trying anyway. It will take a while, maybe even three or four weeks, before you can follow these easy speed contacts in your head without writing everything down! Jot down the call, the operator's name, QTH and report, while just listening to the rest.

Using a program such as ZL1AN, you can select various chunks of text, with mixed characters too. Once you have determined your present 100% comfortable hand copying speed using a pen, set your computer program to send clear text at about 5WPM faster than your hand copying speed. Then sit back, and just listen for 30 minutes, twice per day. First you will only be catching a letter every now and then. This will improve and you will find yourself eventually reading the text in your head with no problem. Have a look at a program called RUFZ by DL4MM. This will send the call signs of the world to you at ever increasing speed and you must enter them correctly via the keyboard. This will improve your typing skills too.

Keep it Up

I have tried to cover just about everything I can think of, taking you from the basic principles, through to running CW contests with a computer. I do hope you have found this book helpful, leading you to a very enjoyable facet of our hobby.

Have fun; keep the practice up daily, hold daily CW contacts and you will get there. I am still working on it, but I am probably older than 90% of you!